# The Disa[ppearing]
# Hedg[ehog]

# Toni Bunnell

www.tonibunnell.com

## Praise for *The Disappearing Hedgehog*

'Brilliant. If you are seriously interested in saving these amazing, interesting little guys this book is a must. After rescuing, caring and watching over local hedgehogs since the end of the 2nd World war you would think I had nothing to learn. And so did I; that was until I purchased Toni Bunnell's book 'The Disappearing Hedgehog'. What a delightfully thoroughly enjoyable little book. A really great read to keep handy close by for the advice.' *Gloria Muir, Hogles Wood Hedgehog Home, rehabilitation centre*

'This excellent, clearly written and authoritative book gives precise and detailed information of how to make our gardens more hedgehog friendly, and how to help and look after hedgehogs that are poorly or who have had bad encounters with dogs, foxes, or garden machinery. Very well presented and illustrated. I would thoroughly recommend this book to anyone with an interest in hedgehogs - and buy one for a friend who should be interested but perhaps is not!'
*Westfield, hedgehog enthusiast*

'I've read my book from cover to cover. Informative and well written.'
*Louise Godden, hedgehog enthusiast*

'I highly recommend it.'
*Geraldine Williams, Veterinary Nurse*

'Having 20 years' experience of hedgehog rehabilitation it's good to know the findings and experiences of other rehabilitators. A good read for both the novice and experienced hedgehog carer.'
*Leighog*

'An excellent book, well presented and factually correct. As an experienced hedgehog carer, I would thoroughly recommend it.'
*Susan Haggas, hedgehog rehabilitator*

'Excellent little book for anyone with an interest in hedgehogs. Simple and straightforward to read, it contains all the facts you need to know about hedgehogs, plenty of advice and informative illustrations. There are case studies of hedgehogs with a variety of illnesses/injuries from Toni's rescue centre, giving a good overall idea of the problems they can encounter. There is even a test quiz at the end so you can review what you have learned. Handy book to read again and again!'
*Amazon customer*

'A very straightforward, informative little book. Worth buying to refer to again and again.'
*Dawn Higham*

A sequel to The Disappearing Hedgehog book has been published and includes the findings from Toni's published research.

Rescuing the Disappearing Hedgehog is available from: www.tonibunnell.com

*Also by Toni Bunnell*

Music Makes a Difference[1]

The Room Between the Floorboards[2]

A Door in Time[3]

The Fidgit[2]

Tales of Sweeper Joe the Hero[3]

Samuel and the Stolen Words[2]

Dreaming While Awake[3]

Into Oblivion[3]

The Nameless Children[2]

A Life Well Lived[2]

Rescuing the Disappearing Hedgehog[1]

Code Switch[3]

\*\*\*

Available as paperback only[1]

Paperback and ebook[2]

Ebook[3]

www.tonibunnell.com

# Contents

| | |
|---|---|
| Introduction | 1 |
| Ecology and Behaviour | 3 |
| What to do when you find a hedgehog | 7 |
| Exceptional Success Stories | 23 |
| Tales from York Hedgehog Rescue Centre | 45 |
| Helping hedgehogs in your garden | 51 |
| Useful tips | 52 |
| Test your knowledge | 61 |
| Publications | 64 |
| Acknowledgements | 66 |
| The Author | 68 |

First published in 2014 by Toni Bunnell, York, UK

First edition printed February 2014
Second edition printed August 2014
Third edition printed November 2016

ISBN: 978 - 1 - 78280 - 236 - 5

All rights reserved. No part of this publication may be reproduced, stored in or introduced into a retrieval system, or transmitted in any form, by any means (electronic, mechanical, photocopying, recording or otherwise) without prior written permission from the copyright holder Toni Bunnell.

A CIP record for this book is available from the British Library.

Printed and bound in Great Britain by
Short Run Press Ltd Exeter

This book is sold subject to the condition that it shall not, by way of trade or otherwise, be lent, resold, hired out, or otherwise circulated without the author's prior consent in any form of binding or cover other than that in which it is published and without a similar condition, including this condition, being imposed on the subsequent purchaser.

Cover/text design and photographs by Toni Bunnell

www.tonibunnell.com

# Introduction

The European Hedgehog, native to the UK for thousands of years, is fast disappearing from our shores. In the past 50 years numbers are estimated to have fallen from 32 million to fewer than one million. The reasons for this are mostly due to habitat loss and other problems caused by humans. This book explores some of the ways in which we, as individuals or in groups, can help to redress the balance.

As Margaret Mead, the anthropologist said: "Never doubt that a small group of thoughtful, committed citizens can change the world; indeed, it's the only thing that ever has."

If each of us adopts this approach we can save the hedgehog and halt the rapid decline in population numbers. We can make sure that the species will continue to be seen by generations to come and ensure that the hedgehog, which was voted as the most iconic animal and the best natural emblem by the British public in 2013, will flourish in our gardens and woodland.

In the following pages you will find ways to encourage hedgehogs to visit your garden, what to do if you find a hedgehog in need of help, some interesting facts and

Tales from York Hedgehog Rescue Centre, as seen through the eyes of Roger, a hedgehog who became a long-term resident. Exceptional success stories, describing the treatment and recovery of seven particular hedgehogs that needed help, are recounted. If you wish you can test your knowledge of hedgehogs both before and after you have read the book.

It is hoped that this book will help to raise the profile of hedgehogs and increase the level of understanding of their needs. In so doing we should be able to halt the disappearance of the hedgehog from the UK.

# Ecology and Behaviour

The minimum weight for a hedgehog to survive hibernation is 650+g, assuming the hedgehog also has a rounded end. It is the weight/size relationship that is important rather than the weight per se. All healthy hedgehogs should, essentially, be almost sphere-shaped.

Hedgehogs do not hibernate continuously throughout the winter months; neither do individual animals necessarily hibernate at the same time. This behaviour is known as periodic arousal.

Triggers for hibernation include decreasing day length and temperature.

Very few hedgehogs carry fleas. Even very sick animals will not necessarily be infested. The hedgehog flea is specific to hedgehogs and will not infest domestic animals such as cats, dogs or humans.

In the UK hedgehogs can have two litters each year. The first litter is born in May or June while the second litter from August onwards.

In the UK hedgehogs can have between two and eight babies (13 have been cited) in a litter but typically have between four and five.

Baby hedgehogs are known as hoglets.

More male hedgehogs are born than females. The ratio is 1.5 males:1 female (Bunnell T. 2001).

Hedgehogs born to second (late) litters gain weight more rapidly than those born to first (early) litters (Bunnell T. 2009).

Male

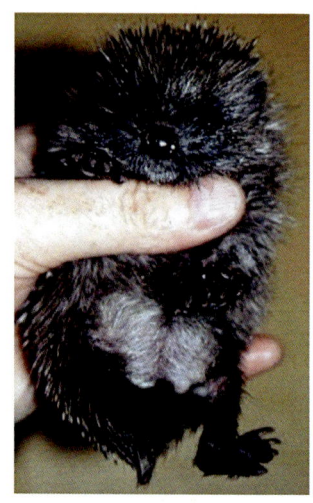

Female

A baby hedgehog's eyes don't open until 14 days after birth.

Birds' eggs constitute 10% of a hedgehog's natural diet.

Contrary to popular opinion the main food source of a hedgehog is not slugs and snails. These make up only 5% of their diet.

The main food source of a hedgehog is beetles (30%).

Caterpillars make up 25% of the diet.

Earthworms make up 11% of the diet.

Hedgehogs cannot climb trees as they do not have retractile claws. However they are very adept at climbing ivy.

Hedgehogs can swim on the surface of water for a limited time and so are able to cross rivers.

When a hedgehog is tightly curled, during hibernation, the layer of air trapped between its spines serves as insulation against the cold by reducing heat loss.

After mating the male hedgehog leaves the female and does not help to look after the young.

Hedgehogs have been reported to live up to seven years in the wild but they normally survive in the wild for only 18 months.

Hedgehogs do not burrow but can dig out a depression in the soil, meaning that they lie relatively flat with the surface of the soil under a covering of grass or leaves.

When rain begins to fall hedgehogs will head for cover.

Apart from during courtship adult hedgehogs rarely vocalise, even when experiencing intense pain. A screaming sound is sometimes emitted when a hedgehog is extremely scared or in pain, but this is a rare occurrence.

Baby hedgehogs emit a 'peeping' sound similar to a baby bird when they are in distress or have been left without food for many hours, usually due to the death of the mother.

The name 'hedgehog' first came into use around the year 1450.

An old name for the hedgehog is 'urchin'.

The eyesight of hedgehogs is poor but they have been reported to be able to recognise a skyline from the outline of trees against the sky.

The main senses of a hedgehog are hearing and smell.

In the wild adult hedgehogs, in their second year, may reach 1300g.

Hedgehogs are not, strictly speaking, territorial but will fend off other hedgehogs from food sources.

# What to do when you find a hedgehog

## Here are some reasons for seeking help:

A hedgehog is found out in the day i.e. any time other than dusk/dawn/night

N.B. Of hedgehogs found out in the day the greater majority are sick or injured. Possibly 1% might be starving and searching for food.

Appears to be injured

Found lying by roadside

Staggering about

During the breeding season (May to October) be aware that an adult female, with babies, might be out in the day foraging or even relocating her babies to another nest. Check the area for a nest before taking in any adult hedgehog during the breeding season.

If you find a baby out of its nest check to see that there aren't any more as usually where there is one there will be more….

Where there is one baby hedgehog there will usually be more

Up to eight babies from the same litter have been found without a mother

Often the mother has died or an animal such as a dog or cat has disturbed the nest

At the first opportunity you should seek advice from a specialist for any hedgehog found. Until then here are some things you can do to help. N.B. All treatment described in this book was prescribed by a vet.

## Temporary housing

A recycle box is ideal. Make sure that the hedgehog can't climb out. Many a hedgehog has had to be retrieved from under the floorboards or from behind a cupboard… Place newspaper in the bottom and provide a towel (with no hanging edges that hedgehogs can get their head caught in, causing strangulation) or hay. If the hedgehog feels cold to the touch provide a hot water bottle wrapped in a towel and place under or beside the hedgehog. **Never** place a collapsed hedgehog on top of a heat source from which it will be unable to move away. Hedgehogs are unable to sweat and can quickly overheat and die.

# Feeding

Do not give large first meal
Feed little and often
Baby: Kitten biscuits and/or cat meat (no gravy) and goat's milk. Lactose-free milk is also available in most supermarkets. **Never** give normal cow's milk as hedgehogs are lactose intolerant and it will make them ill. If the hedgehog is reluctant to eat try warm scrambled egg (no milk). A glass lemon squeezer is an ideal way to feed several babies at once and helps teach them to lap on their own (*see photo*).
Adult: Kitten biscuits and/or cat meat (without gravy). Provide a heavy, shallow dish for water that cannot be easily tipped over. Charity shops are a good source of these.

## How can you be sure that the hedgehog is eating?

A hedgehog can often appear to have eaten while in fact it might just have walked in the food and distributed it round the container (*see photo*).

Only by **weighing daily** can you be sure that a hedgehog is eating and hence gaining weight. You should weigh it on arrival. It should gain about 10g a day (see photo on page 11).

Hoglet drinking Esbilac from lemon squeezer

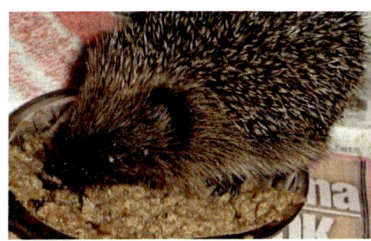
Hedgehog sitting in dish

Digital scales

## Is the hedgehog underweight?

A hedgehog should have a very rounded rear end. A tapered rear end indicates that it is thin and underweight for its size. The thin hedgehog shown weighs 495g but needs to be over 900g for its size. Weight-size relationship is all important (*see photo*).

## How to tell if a hedgehog is hibernating

A hedgehog that is hibernating will be very tightly curled and might even appear to be dead. There will normally be no movement of the spines and the head and nose will not be visible (*see photos*).

## Flystrike

Flies sometimes lay their eggs on sick/immobile hedgehogs. The eggs appear as tiny grains of rice (*see photo*).

It is important to remove as many fly eggs as possible using an old toothbrush or tweezers. Once maggots hatch out they must be removed immediately as, if left, they will enter the hedgehog wherever possible, particularly in areas where there are open wounds. They can be removed easily using tweezers.

Very thin: tapered end        Healthy: rounded end

Not hibernating        Hibernating

## Ticks

Ticks are a sign that a hedgehog is sick (Bunnell T. 2011). A hedgehog can become seriously anaemic if infested by many ticks (*see photo*). The ticks can easily be removed using curved-ended forceps or tweezers. Grip the tick near the head end, where it is attached to the hedgehog, and pull quickly in an anti-clockwise direction, taking care that no spines are trapped in the forceps. Removing the tick in an anti-clockwise direction causes less discomfort for the hedgehog as the tick attaches itself to a hedgehog using a corkscrew motion in a clock-wise direction. It is not advisable to spray ticks using products licensed for cat and dog use, as these can often have detrimental effects on the hedgehog, especially one that has breathing difficulties.

## Fleas

Contrary to popular opinion very few hedgehogs in the UK carry fleas. Where they do this is often indicative of a sick animal. The flea is specific to hedgehogs and will not infest dogs, cats or humans. Treatment for fleas is any flea powder that is licensed for use on exotic birds such as canaries. Sprinkle some powder lightly and sparingly over the back of the hedgehog, avoiding the head.

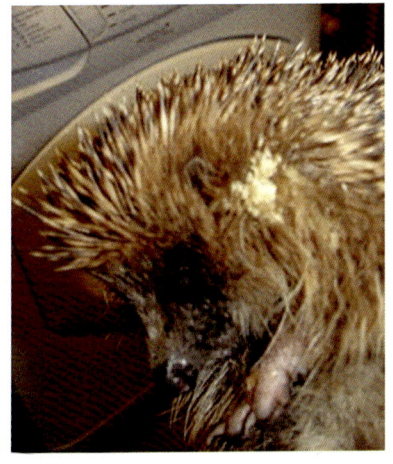

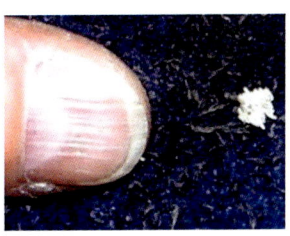

Fly eggs appear as tiny grains of rice on this sick hedgehog

Curved-ended forceps for tick removal seen here with flea

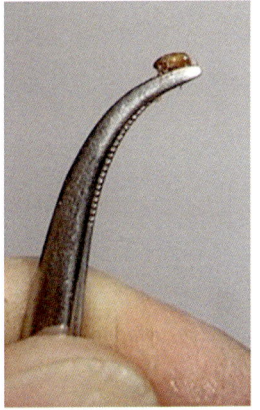

A hedgehog infested with ticks indicates that it is sick/injured

## Ringworm/Mange

Ringworm is a fungal skin condition that often begins on the nose. Mange is caused by a microscopic mite that lies just beneath the skin surface (*see photo*). Both ringworm and mange can be treated using Thursday Plantation Tea Tree Antiseptic Cream applied sparingly to affected areas. Neat Tea Tree oil must not be used in place of the cream. Neem Oil can also be used to treat both these skin conditions, to good effect.

Severe mange or ringworm can result in very heavy spine loss which makes the affected hedgehog very vulnerable to predators (*see photo*).

## Behaviour

Hedgehogs that are very active are often thought, mistakenly, to be fit and well. This is not usually the case. A hyperactive hedgehog that will not settle and sleep, particularly during the day, usually means that it is poorly and in much discomfort. This can be caused by many things such as a heavy parasite burden in the gut, a bacterial gut infection, an injury or pain.

Hedgehogs rarely vocalise, even when experiencing intense pain, hence a lack of noise does not equate to no pain.

Ringworm usually starts on the nose

Severe spine loss left this young hedgehog very vulnerable to predators. Treatment with Tea Tree Antiseptic Cream, used intermittently with Neem oil, resulted in new spine growth all over. Amos was released back to the wild with a full set of spines.

This lack of vocalisation probably has a survival function for the hedgehog as a predator will not be attracted to an injured animal that is unable to retreat or curl into a tight ball, its only method of protection.

## Broken leg?

If a hedgehog is seen to be dragging a leg, this indicates that the leg is injured and possibly broken. Apart from an unbalanced gait when walking, another way to tell that a hedgehog has a broken leg is that, when it curls into a ball, one leg remains sticking out.

It is important to establish if there is a wound associated with the break and also, if there is any infection. Hence attention from a vet must be sought at the earliest opportunity.

**Dog or fox attacks** are normally responsible for broken legs, especially hind legs. Checking that there is no hedgehog in your garden, before letting your dog out at night, will save countless injuries and hedgehog deaths every year.

This hedgehog has a broken front left leg that it is unable to draw inside when it tries to curl up. No infection was present and the leg was splinted, allowing it to heal.

Gardening gloves are ideal for holding and examining hedgehogs

Rubber car mat offers waterproof protection for wooden house

## Wobbling/shaking hedgehog

A hedgehog that is wobbling or shaking is most likely to be dehydrated/cold/poisoned/or have low blood sugar. It will need to see a carer or vet as soon as possible. If you are an experienced carer you can help by making up a rehydration solution. Mix a tablespoon of sugar with a teaspoon of salt in a litre of water. This can be frozen without losing its properties. The solution can be administered using a plastic syringe or dropper. Care must be taken to give only one drop at a time while holding the hedgehog forward slightly so that the liquid does not enter its airways which could choke it (*see photo*).

## When a hedgehog refuses to eat
### Syringe or dropper– feeding

If you have an adult hedgehog that won't eat, or a baby hedgehog that is too tiny to feed itself, it is very important to keep it hydrated using a 1ml syringe.

## Getting a hedgehog to unroll

It is often necessary to check for injuries and, in order to do this, the hedgehog needs to be encouraged to uncurl. Holding the hedgehog in one gloved hand, while stroking the back of the hedgehog with the other gloved hand, should produce the desired effect.

Hoglets being fed using a 1 ml syringe

## Visit to vet/wildlife rescue centre

After being seen by a specialist, on occasion a hedgehog will be found to have no obvious injuries or conditions in need of treatment. In these circumstances it is very important to always release it immediately back where you found it. The only reasons for not doing this would be the presence, close to the finding site, of a known badger sett, main road, building site or garden where an aggressive dog lives. In this case it is advisable to identify a suitable alternative safe site for releasing the hedgehog.

It is also advisable to release all hedgehogs that have been restored to full health, back to where they were found. Hedgehogs know their area and will easily find their nest again. I have first-hand experience of their ability to locate their last nesting place with ease.

# Exceptional Success Stories from York Hedgehog Rescue Centre

Over the course of the past 26 years many hedgehogs have arrived at the doors of York Hedgehog Rescue Centre. Several are notable in terms of the degree of injury sustained or the unusual nature of their predicament.

From the multitudes I have chosen seven in particular that lend themselves to a closer look at what ailed them, how they were treated and the outcome. All these hedgehogs received specialist care and all treatment was prescribed by a vet.

| | |
|---|---|
| **Stephen** | 24 |
| **Leon** | 27 |
| **Hazel** | 30 |
| **Snow & Sun** | 32 |
| **Bess & babies** | 36 |
| **Harry** | 39 |
| **Bradley** | 42 |

# Stephen

On June 26, 2008, five tiny hoglets arrived at York RSPCA animal home. They had been found in a garden in York where their mother lay dead. They were in urgent need of help and I took them home to look after them. They weighed between 109 g and 146 g.

The smallest one behaved differently to the others and hung back when food was put out. On closer examination I found that he had a slight depression on top of his head suggesting that something had landed on him, causing a head injury (crush injury). This was confirmed when he suddenly started to fit.

Stephen, as I called him, was unable to reach his food as he was unable to move in a co-ordinated fashion. I hand-fed him by putting small amounts of puppy food and scrambled egg into his mouth which he ate easily. I continued until he was able to feed himself.

**Treatment**: 1 week course of Baytril, oral antibiotic

Stephen (far right) was unable to reach the food by himself as he could not walk in a co-ordinated fashion

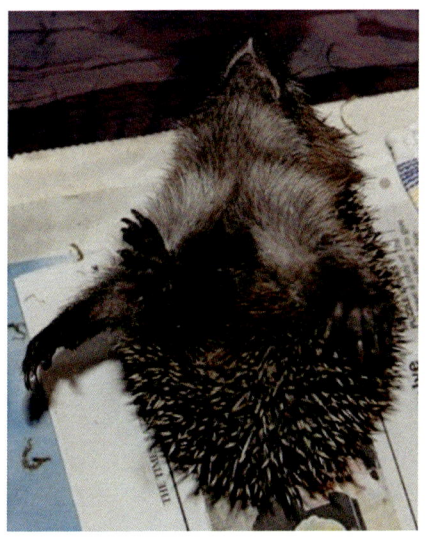

Stephen fitting, shortly after arrival

Stephen responded well to treatment and soon began to eat on his own

Stephen made a complete recovery. On July 31, 2008, a dry, mild day, he was released at 477 g. His spines were marked with a small amount of emulsion paint to allow easy recognition. He was released to an open garden with a choice of hedgehog houses, and food and water provided. All sightings were recorded by the house owner and he continued to appear throughout the summer.

# Leon

Leon had a very close call. He was found early one Monday morning on August 1, 2012, in a newly built concrete trench from which there was no escape. The person who found him did not realise how poorly he was. They removed him from the trench and Leon made his way slowly back towards his nest on a grassy bank in an area of York where I am monitoring the hedgehog population. He was found by Martin, a gardener, who promptly phoned me saying that the hedgehog looked almost dead.

I arrived to pick up Leon within 15 minutes of the phone call and was shocked by his appearance. It seemed more than likely that he had been trapped in the concrete trench since Friday evening as this was the last time that workmen were in the area. He was extremely thin, his claws were completely worn down and his toes were bleeding with the effort of trying to escape from his prison.

Leon was now collapsed and hardly moving. Martin had found him at the bottom of the grassy bank. This meant that Leon had dragged himself two hundred yards from where he had been taken out of the concrete trench, to the bank where his nest was. He had then been too weak to manage the last few yards. As well as being collapsed, Leon was very thin, very cold, starving and covered in fly eggs. Fortunately, help was now on hand.

I took him home and removed the fly eggs. I then gave him 10 ml subcutaneous fluids, having been shown how to do this by a vet. I also gave him some special milk substitute, Esbilac, using a 1 ml syringe and he took 12 ml. Leon also received a course of antibiotics as he had injured his toes, and a wormer for lungworm.

**Treatment**:
1 week course of Baytril, an oral antibiotic
1 week course of Telmin (for lungworm)

Initially Leon weighed 508 g on arrival, and was emaciated, being far too thin for his size. Twelve days later, on August 13, he had gained a total of 224 g. He now looked like he should look, almost spherical when fully curled into a ball.

On August 13, 2012, I released Leon back to the grassy bank where he had been found by Martin and watched as he walked up the bank to his nest. I gave him a distinctive emulsion mark before releasing him. For some considerable time after his release I was able to see him on footage obtained using remote Infra-Red cameras. Against all the odds Leon had made a miraculous recovery and was now back in an area that he knew.

Leon on arrival: very thin, with elongated rear end

Leon after 12 days of care at York Hedgehog Rescue Centre

Leon being released. N.B. A ramp has been built to prevent further entrapments (see photo page 59)

## Hazel

On July 22, 2006, Hazel was admitted as an emergency. She had been the victim of a savage dog attack which had left her with serious head injuries including damage to her left eye socket. She was immediately taken to the Minster Veterinary Practice, York, where an operation was carried out. It was not possible to save her eye.

Two days after the operation, which involved using a general anaesthetic, Hazel gave birth to four babies, all of whom were born dead. This was a very sad outcome to what might have been another uneventful summer for this otherwise healthy adult female hedgehog and her four offspring.

I brought Hazel to my enclosed garden where she recovered steadily. It was not possible to release her back to the wild as it quickly became apparent that she had suffered brain damage as a result of the dog attack. She was unable to forage for herself but was otherwise able to lead a normal life.

**Treatment**: 1 week of Baytril, an oral antibiotic

Hazel was put into one of the open hedgehog houses in my garden but chose to build her own nest elsewhere. She continued to live happily in my garden for the next two years. However, there is no escaping the fact that, had she not been savaged by a dog, she could have lived these years in the wild and had more babies.

Hazel after being savaged by a dog

Hazel with wound fully healed

## Snow and Sun

On July 11, 2012, I received a phone call from York RSPCA Animal Home. One of the RSPCA inspectors had brought in a juvenile hedgehog with a difference. This one was white and a true albino. In all the years that I had been doing hedgehog rescue I had never once seen an albino hedgehog. I had seen a few blonde hedgehogs with black eyes and pale colouration, but none that were albino with red eyes.

I contacted the finder to say that I was now looking after the albino hedgehog that he had found. Two days later I received a call from him to say that he had found another in his garden in the same place. Amazingly this one was also albino and turned out to be the sister of the first one. I collected her from the finder, named her Sun and put her with her brother, Snow.

The gene that causes albinism is recessive, meaning that its effects are only seen when two copies of the gene are present in a hedgehog. This means that an albino hedgehog will inherit one copy of the albino gene from each parent. The likelihood of two hedgehogs meeting, that both possess the albino gene, is rare, hence resulting in few sightings. When both parents possess the gene for albinism there is a 1 in 4 probability that they will have a baby that is albino.

Both Snow and Sun were found outside during the day indicating that they were in need of care. Snow weighed 325 g on arrival and Sun, his sister, 345 g, when she was collected two days later. During the course of their stay with me, at York Hedgehog Rescue Centre, they required treatment for intestinal worm infestations including nematodes and lung worm.

**Treatment:**
1 week course of Telmin (granules) for lung worm
1 spot of Profender Spot-on (cats) for nematodes

On August 7, 2012, Snow and Sun were released to the wild. Snow weighed 727 g and Sun weighed 563 g. They were released to an open, but protected area that I am monitoring on a long-term basis using remote Infra-Red cameras. Footage from the cameras showed regular appearances by both Snow and Sun, with Snow appearing more often at the feeding station. Hedgehog houses are in the area and food and water available at all times, in addition to natural food.

On October 7, 2012, Snow no longer appeared on the cameras while Sun was around until mid-November. There was no further sign of Snow until April 9, 2013, when he appeared at the feeding station and most nights from then on. Sun appeared on no cameras during 2013 but was sighted on a regular basis by a

Snow and Sun: poorly camouflaged against the grass

Snow with normal coloured hedgehog

Pip, a blonde hedgehog, showing his darker spines and black eyes compared with albino, Snow

local resident. Apparently both Snow and Sun appear in the evening at similar times but from different directions. Both survived a second winter and have appeared regularly on camera footage. In July 2014, Sun was seen out in the day and I brought her home to monitor. She seemed fine and gained weight, only to give birth to two babies three days later. Her return to the wild with her babies is imminent.

This is a remarkable success story for two hedgehogs who might have been expected to have a limited prospect of survival. Clearly, their reduced camouflage has not rendered them prey to the foxes that frequent the area and they are living a life in the wild – free.

Snow, 2012                    Sun, 2014

## Bess and her babies

On July 1, 2008, a phone call alerted me to an adult female hedgehog that was dragging her legs and out in the day in a garden in the middle of York. I called to collect her on the way home from work. The finder said that the hedgehog normally lived in a compost heap which intuition told me to check. I looked through the nest of leaves that the hedgehog had built and found three small hoglets curled up at the bottom. I promptly gathered them up, phoned the Minster Veterinary Practice to see if there was a vet who could have a look at the mother, and drove the family there.

The vet on duty examined the adult hedgehog thoroughly and concluded that no bones were broken. He thought that the most likely reason for the immobile back legs was nerve damage in the pelvic region possibly caused during giving birth to the babies. He gave her a steroid injection to help reduce any inflammation and declared her otherwise healthy.

I took Bess, as I named her, home and put her, with her three babies, in one of the hedgehog houses in my garden. The best course of action, when taking in a mother hedgehog with babies, is to place them in as natural environment as possible and leave them to it. Any interference may result in the mother abandoning the babies and, in cases where the babies are very young, possibly eating them. The babies weighed 87, 94 and 94 g and seemed to be feeding well from Bess. There were two females and one male. I checked and weighed the babies from time to time to make sure that Bess was feeding them successfully.

Seventeen days after Bess had been discovered, vulnerable and struggling, she started to display some movement in her hind legs. Up until then she had been getting round the garden trailing her legs behind her.

Bess made a complete recovery and was returned to the garden where she was found. The lesson to be learned from this experience is always to check the garden where an adult female hedgehog has been found just in case there are any babies tucked away in a nest somewhere.

Bess on arrival, with her legs trailing behind her, unable to walk properly

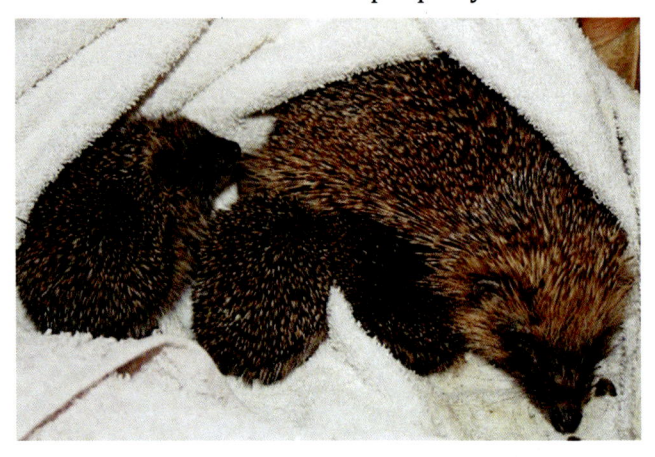

Bess with her babies, before being moved to a hedgehog house in my garden

# Harry

Harry came to me via York RSPCA Animal Home on June 18, 2010. He had been found in a puddle on the banks of the River Ouse, York. Harry was very cold and had been attacked by a domestic animal such as a cat, judging by the size of the tooth marks. There were eight puncture wounds visible, each of which was infested with maggots. Flies had been attracted to the injured hoglet and had laid their eggs on his wounds. Fly eggs usually hatch in less than 24 hours when conditions are warm and humid.

Harry weighed 121g on arrival and his eyes were closed. A hoglet's eyes open when it is about 14 days old so this meant that Harry was younger than this. He was given a course of oral antibiotics to help his wounds heal. I removed all the eggs and maggots from the wounds and a special honey ointment was later applied. This is a recommended treatment prescribed by the vet. It works by effectively removing the oxygen needed for aerobic bacteria to grow.

As Harry was so tiny he was unable to pass urine or poo on his own. A cotton wool bud covered in Vaseline was used to help him wee and poo. When his bottom became sore a small amount of Vaseline was applied to soothe it.

Harry was fed Esbilac, a special milk substitute, using a 1ml syringe. At first he took only a few ml at a time but this soon increased until he was taking up to 13ml in one sitting. He gradually progressed to eating cat meat and by July 7 had reached 313 g.

**Treatment:** 1 week course of Baytril, oral antibiotic

Harry made a full recovery but sadly lost the use of his right eye which had been destroyed by maggots before he was found. Hedgehogs rely mainly on the senses of smell and hearing to survive, rather than sight. However, they have been found to be able to locate particular surroundings at night by being able to see the outlines of trees. In Harry's case the loss of one eye would not prove a problem for surviving in the wild.

On July 22, 2010, Harry weighed 495 g and was well and healthy. As it was the middle of summer, and the weather was dry and mild, Harry was released to an open garden in a village on the outskirts of York. He was marked with a unique mark using a small amount of emulsion paint on his spines. The person who offered her garden for Harry to live in also provided a hedgehog house and food and water on a daily basis. From time to time he was caught, weighed to check he was fine, and re-marked. He was last seen in June

2013, before the resident moved house. The new house owner has promised to look out for Harry.

Harry on arrival: showing puncture wounds and fly eggs

Harry fully recovered

Bradley tapered      Bradley fully recovered

Bradley about to be released

# Bradley

This is Bradley on arrival 010414.

Just a little chap weighing in at 313g.

Wendy found him wandering the streets of York on her way to work and took him with her. He was a little thin and obviously poorly/starving as he was out in the day. He must have been born very late last year (2013) and has managed to survive all winter which, fortunately, has been very mild. Bradley was placed in my kitchen for observation.

Bradley held his own and I kept him hydrated using lactose-free milk and also Esbilac, a puppy milk substitute without lactose. Following his arrival he did not show much interest in food, other than the occasional meal worm, so I offered him lightly scrambled egg (without milk). Eggs constitute 10% of the diet of wild hedgehogs so are fine in moderation. He loved it and ate it all.

It was not all plain sailing, however, as Bradley became far from well, showing the tapered end sadly typical of a poorly hedgehog. The vet gave him an injection of Levacide, for lungworm, Baytril, an antibiotic, and subcutaneous fluids to counteract dehydration. He was quite hyperactive at times and I syringe-fed him to keep him hydrated.

On April 20, Bradley weighed 513g (a gain of 212g in a week) and could no longer be considered to be a disappearing hedgehog.

Bradley was released back to the wild on April 28 weighing 721g.

He was picked up by the remote cameras that I use to monitor the area, on May 2, 2014.

# Tales from York Hedgehog Sanctuary

**Roger** came into my care in October 2009. He was an adult male and was suffering with pneumonia. An indentation in the spines on top of his head suggested a conflict with a grass cutting device such as a strimmer. A strimmer not only cuts through spines and soft tissue but can also land heavily on the head of a hedgehog and cause brain damage. This is what seems to have happened to Roger. In common with many brain-damaged hedgehogs Roger was unafraid of humans.

Over time he responded to my voice to the extent that he would come running down the garden when he heard his name called indicating that food had been put out. Roger was unable to fend for himself and hence return to the wild was not possible. He spent the remaining 3 and half years of his life living free in my enclosed garden, as recounted in these tales.

## Food on Demand

In the ensuing days Roger was strongly encouraged to feed himself. But he was having none of it. Why bother to feed yourself when someone else will do it for you was his motto. No morsel of scrambled egg would he let touch his lips. It was hard-going looking after Roger.

Then one day, without a word of warning, Roger decided that after all he might like egg. Once he'd started eating there was no stopping him. He wolfed it down and put on weight at speed. Roger was on the road to recovery.

## Roger goes off his food

Roger stopped eating. At the turn of midnight he lost all interest in food. The temperature fell below zero in the night, the water in the bird bath had frozen solid and Roger had pneumonia - again!

Before there was time for self-pity to kick in he was moved to the Quail brooder in the kitchen, faster than the blink of an eye, faster than a Roe deer taking flight and certainly faster than Roger moved on an average day. Now in intensive care his spirits rallied and he saw a ray of hope on the horizon.

## Roger Fights Back

Roger was not going to be beaten easily. Life was too good and after all Christmas was just round the corner and he was eagerly anticipating his presents.

He seemed to be being monitored round the clock judging from the number of times his towel was lifted up and he was checked. As if by a miracle, after the unbelievably short time of 3 days, he started to make a recovery and bounced back into the land of the living. This time he bypassed the scrambled egg and went straight for the pedigree chum puppy meat. He still wasn't feeling a hundred per cent though and was being topped up with Esbilac, a special milk substitute. Things were looking good again for Roger.

## Roger Becomes a Star

Roger couldn't quite believe his meteoric rise to fame. It was hard to take in for a humble hedgehog such as himself. One minute he was going about his normal humdrum existence - sit and wait each morning for breakfast, his house to be cleaned, his bed to be made up nice and fresh, just as he liked it, supper laid on in the evening.... in a word all his needs were met and he wanted for nothing - the next minute, without a word of warning, he was propelled into the limelight. He had always wondered what was meant by 'the limelight' and now he knew.

On November 25, 2009, he was abruptly woken from his morning slumber to find a television camera aimed directly at him. But Roger was not a hedgehog that was fazed easily. He took it all in his stride and posed for the camera, taking care to always show his good side. Having performed for the camera he went back to his bed and slept for the rest of the day. Little did he know that he had been catapulted into the stratosphere of the world of fame, and had done his bit to highlight the plight of hedgehogs in Britain today.

All throughout Yorkshire people watching Look North on their televisions that evening commented on how calm and collected Roger had appeared for what was actually his debut in the world of television. Roger had become a star!!

## Roger takes up gardening

Think not of what Britain can do for the hedgehog but of what you, the hedgehog, can do for Britain. With this edict firmly in his mind Roger decided to take up gardening. With winter nights spent walking the bathroom tiles he eyed up his surroundings. The large potted plant in the corner looked like a good place to start. Climbing up to soil level took some doing and he paused at the top to get his breath back. He then set to with vigour and enthusiasm digging out the soil and distributing it evenly round the bathroom. Hedgehogs were never meant to be surrounded by cleanliness, he thought to himself, and today he had proved his point. The bathroom floor now closely resembled an allotment and, if Roger had his way, would stay like this for the foreseeable future.

Roger takes up gardening

Roger looks forward to spring

# Helping Hedgehogs in your garden

Useful Acronym: **OPALS**

**Out in the day:** a hedgehog is most likely to be poorly and in need of help.

**Ponds:** need to have several places where a hedgehog can climb out safely.

**Access:** leave a few areas where hedgehogs can get into your garden from other gardens to feed.

**Leave** an area undisturbed in your garden where a hedgehog can build a nest.

**Strimmers:** always check any area before using a grass-cutting device such as a strimmer, in case a hedgehog is asleep in the grass.

# Helping hedgehogs in your garden

You might already have hedgehogs in your garden. One way to find out is to check for the presence of foraging holes in the grass.

## Providing a house

A perfect house can be made by placing a paving stone over four bricks (one under each corner) and pushing hay or dried leaves underneath. The ideal place for a house is under a hedge or large shrub, behind a shed or in a quiet shady corner, but never in direct sunlight. An old rubber car mat can be placed over the entrance to keep out drafts (*see photo* page 19).

## Feeding station / Food

An excellent feeding station can be made by using an upside down plastic cat basket or, better still, two. If using one basket place a brick on top to prevent it being overturned by another animal and a brick in the entrance to prevent a cat, for instance, from gaining access to the food. Food can be placed in dishes which helps prevent dried food from going mouldy. If using two baskets place one on top of the other, leaving a gap of a few inches for the hedgehog to crawl underneath to the food (*see photos*). Cat biscuits or dried mealworms are ideal food for a hedgehog.

Foraging holes in the grass indicate the presence of hedgehogs in your garden

Feeding station on my allotment

Two cat baskets provide a feeding station that hedgehogs can easily access but cats cannot

## Water

Water is essential for hedgehogs to survive, particularly during the summer months. A large frost-proof dish can be obtained cheaply from a charity shop. A plastic plant pot base will also work well.

## Ponds

A pond in your garden is a great source of water but the utmost care must be taken to make sure that any hedgehog that might chance to fall in will be able to get out easily and quickly. I am often asked why hedgehogs drown in ponds as it is known that they can swim, at least on the surface. The answer is simple:

**Hedgehogs can swim but they can't swim forever**

They will eventually become exhausted and drown. Any ramp that is positioned at the side must have a rough surface, such as roofing felt, to prevent it from becoming slippy. Hedgehogs do not have retractile claws like a cat and, as such, cannot get a good purchase on wet wood.

Ideally, a pond should have shallow sides with shale or small stones and rocks on the edge. This will enable the hedgehog to swim to the edge and climb out easily.

Plastic plant pot base collects rain water on an allotment

The pond shown here was made from an old sink submerged in the ground. Stone is placed in pond near edge as stepping out point.

An ideal pond with gravel at the shallow edges so a hedgehog can easily climb out.

Another option is to lay sacking round the edge of a pond, so that the hedgehog can climb out wherever it chooses. I have seen instances where sacking has been positioned in a couple of places round the edge of a large pond and hedgehogs have still drowned. The hedgehog is not aware that the ramp or sacking is on the other side of the pond and tries to get out at the point where it has found itself. Sadly, all its efforts are to no avail and it will eventually perish.

## Ramps

After repeated instances of hedgehogs falling into an enclosed concrete trench, near a church, the caretakers offered to build a ramp to act as an escape route. This has worked brilliantly.

## Access

A hole four inches high and six inches wide is sufficient to allow a hedgehog to get under a fence. The more access points you have the more easily hedgehogs will be able to make their way to and from your garden to others in the locality.

## Leaving an area of garden undisturbed

Hedgehogs need a remarkably small area in which to make a nest and rear their young. In an allotment, close to my allotment, a mother hedgehog gave birth to, and successfully raised, five babies in an area of uncut grass no more than four feet by four feet. You can also help by putting grass cuttings and dried leaves in a heap under any bushes in a shady part of your garden.

# Hazards

**Strimmers** can do untold damage to hedgehogs and cause terrible injuries. This can be avoided by checking any long grass for hedgehog nests that may have babies in them or sleeping hedgehogs. The hedgehog's automatic defence, when faced with danger, is to curl even tighter in a ball and remain still. This is no defence against a strimmer.

**Netting** such as the type used to protect vegetable crops from birds or that used with recycle boxes can be lethal when caught round a hedgehog's spines as it cannot escape and often dies from the resulting wound even if freed.

**Holes in the ground** such as those produced during building development are lethal to hedgehogs as they are unable to climb out.

**Drainpipes** are inviting places to hedgehogs who often climb or fall down them and are unable to escape. If the hedgehog is on its back, and tightly curled, then by wearing a pair of garden gloves and inserting one gloved finger into the underside of the hedgehog you should be able to withdraw your hand from the hole with the hedgehog tightly fastened to it.

**Bonfires**: Ensure no hedgehog is present before lighting.

**Domestic animals loose after dusk** pose a potential threat to any hedgehog wandering through your garden. The number of dog attacks on hedgehogs has arisen dramatically over the past few years. Even a very small dog can exert a jaw pressure sufficient to paralyse a hedgehog, leading to it having to be put to sleep.

**Page 59**: A small gap under gate allows hedgehogs to enter and leave the garden. Hedgehog barely visible in summer nest: susceptible to strimmer injury. Plastic netting can cut like a knife and cause serious injury.

Ramp especially built to allow any hedgehog to escape from the concrete trench

Rat traps pose a terrible threat. Also: slug pellets, pesticides, rat poison

# Useful points

- A hedgehog found lying out in the sun is not sunbathing. It is in urgent need of help.
- If you need to use a strimmer, start from the top of the grass and move down gradually, checking for hedgehogs at ground level. A hedgehog will not run away when it hears the noise of the approaching strimmer but will curl into a ball and needs to be moved away.
- When dismantling a compost heap start from the top and remove a section at a time using a garden fork. Forking near the bottom can cause fatal injuries to a hidden hedgehog.
- Before lighting a bonfire ALWAYS dismantle and move the entire assembly in case hedgehogs are nesting in it.
- An adult female, found during the breeding season, might have young and should be returned to the area, where found, as soon as possible following treatment. A lactating female will have nipples evident on her underside.
- **Remember**:
A hedgehog that is out in the day needs help.

# Test your knowledge

**1.** How many babies does a hedgehog typically have in a litter?
**A**: 2
**B**: 4 to 5
**C**: 8

**2.** How many litters are born to hedgehogs in the UK?
**A**: 2
**B**: 3
**C**: 1

**3.** What is the minimum pre-hibernation weight necessary to survive hibernation, assuming a rounded end for the hedgehog in question?
**A**: 400g
**B**: 800g
**C**: 650g

**4.** What is the main food source for a hedgehog?
**A**: Beetles
**B**: Slugs
**C**: Earthworms

**5.** How long does a hedgehog normally live in the wild in the UK?
**A**: 24 months
**B**: 12 months
**C**: 18 months

**6.** Are hedgehogs territorial?
**A**: No
**B**: Occasionally
**C**: Yes

**7.** An old fashioned name for a hedgehog is:
**A**: Hodgeheg
**B**: Urchin
**C**: Spiky

**8.** When it rains a hedgehog will (typically):
**A**: Curl into a ball
**B**: Run for cover
**C**: Ignore the rain

**9.** How many days after birth does a hedgehog baby's eyes open?
**A**: 10
**B**: 21
**C**: 14

**10.** Baby hedgehogs are known as:
**A**: Kittens
**B**: Spikelets
**C**: Hoglets

**11.** Around which year did the word 'hedgehog' first come into use?
**A**: 1066
**B**: 1800
**C**: 1450

**12.** What is the most poorly developed of the hedgehog's senses?
**A**: Hearing
**B**: Sight
**C**: Smell

**13.** A hedgehog found out in the day is invariably:
**A**: Exploring a new area
**B**: In need of help
**C**: Looking for a mate

**14.** Hedgehogs should never be fed cow's milk as they are intolerant of:
**A**: Lactose
**B**: Nuts
**C**: Gluten

**15.** What is the longest time that a hedgehog has been recorded living in the wild?
**A**:   7 years
**B**: 18 months
**C**:   2 years

**Answers to Test Your Knowledge:** 1:B  2:A  3:C  4:A  5:C  6:A  7:B  8:B  9:C  10:C  11:C  12:B  13:B  14:A  15:A

## Useful contacts:
British Hedgehog Preservation Society:
http://www.britishhedgehogs.org.uk   01584 890801
**Facebook: Help York's Hedgehogs**

## Publications
Bunnell T. 1998. Susceptibility of juvenile hedgehogs to disease: Some observations. Imprint, Newsletter of the Yorkshire Mammal Group, No. 26.

Bunnell T. 2000.Tea Tree Antiseptic Cream: A new treatment for ringworm and sarcoptic mange in the hedgehog *Erinaceus europaeus.* Journal of the American Holistic Veterinary Association, September, Vol.19, No.2: 29-31.

Bunnell T. 2001.Treatment for ringworm and sarcoptic mange in the hedgehog *(Erinaceus europaeus).* The Rehabilitator, British Wildlife Council Newsletter 30: 5, January.

Bunnell T. 2001. An effective, harmless treatment for tick *(Ixodes hexagonus)* infestation in the hedgehog (Erinaceus europaeus). Journal of the American Holistic Veterinary Association, January, Vol 19, No.4: 25-26.

Bunnell T. 2001. The importance of fecal indices in assessing gastrointestinal parasite infestation and bacterial infection in the hedgehog (*Erinaceus europaeus*). Journal of Wildlife Rehabilitation, Vol.24 (2):13-17.

Bunnell T. 2001. The incidence of disease and injury in displaced wild hedgehogs (*Erinaceus europaeus*). Lutra, Vol. 44 (1):3-14.

Bunnell T. 2001. Hedgehog Rehabilitation (Letter). Journal of Wildlife Rehabilitation, Vol. 24 (4):3.

Bunnell T. 2002. Wild Hedgehogs. Mammal News, No. 130:7.

Bunnell T. 2002. The assessment of British hedgehog (*Erinaceus europaeus*) casualties on arrival and determination of optimum release weights using a new index. Journal of Wildlife Rehabilitation, Vol. 25, (4): 11-21.

Bunnell T. 2007. Hedgehogs in trouble. Yorkshire Wildlife Trust magazine: 12-13.

Bunnell T. 2009. Growth rate in early and late litters of the European hedgehog (*Erinaceus europaeus*). Lutra, Vol. 52 (1):15-22.

Toni Bunnell, Kerstin Hanisch, Joerg D. Hardege and Thomas Breithaupt. 2011. The Fecal Odor of Sick Hedgehogs (Erinaceus europaeus) Mediates Olfactory

Attraction of the Tick Ixodes hexagonus. Journal of Chemical Ecology: Volume 37 (4): 340-347.

Speck S., Perseke L., Petney T., Skuballa J., Pfäffle M., Taraschewski H., Bunnell T., Essbauer S., Dobler G. 2013. Detection of Rickettsia helvetica in ticks collected from European hedgehogs (Erinaceus europaeus, Linnaeus, 1758). Ticks and tick-borne diseases 4 (3): 222-226.

Many of these publications can be obtained in full: www.tonibunnell.co.uk

## Current research projects:

These include trying to shed more light on the factors that govern hibernation in hedgehogs.

In addition, I am continuing with a study which involves long-term monitoring of hedgehog populations, using remote Infra-Red cameras, in York.

## Acknowledgements

I would like to thank all the staff at York RSPCA Animal Home, in particular Deputy Manager, Ruth McCabe and Marie Sandle, Sue Heathcote, and Richard Jackson for the invaluable support that they have shown for York hedgehogs in need. Also, for the help and advice they have given me during the 23 years that I have been running York Hedgehog Rescue Centre.

The help and support provided by the staff at the Minster Veterinary Practice, York, led by vets Don McMillan and Mark Goodman, has also been much appreciated. In particular I would like to thank vet Pete Crossan who has treated so many hedgehogs, always with great care and enthusiasm, ensuring the best scenario in every case. The co-operation of the veterinary staff has helped to develop a greater knowledge base regarding how best to treat the hedgehog, a wild animal that does not always present symptoms in an easily accessible manner!

I would also like to thank the growing band of hedgehog carers in York who have helped me greatly in recent times. These include Emma Farley, Jan Collinson, Linda Quantrell and Gill and Phil Barrett.

And - finally – I would like to thank my husband, Paul, for his unwavering and whole-hearted support of all my literary ventures and for listening patiently to yet another finely detailed account of the progress of yet another hedgehog installed in my rescue centre; Also, for agreeing to share a hotel room with little Harry, cited in 'Exceptional Success Stories', who I felt I couldn't leave during a trip to Cumbria.

Thank you to all of you. You have helped to make the journey of discovery, in determining what works best when treating sick and injured hedgehogs, all the more enjoyable and rewarding.

# The Author

Toni Bunnell has had a passion for animals and their welfare since the age of three when she reprimanded a visitor at a zoo for feeding the bears when the sign specifically stated 'Do not feed the bears'. The passion never left her and resulted in her pursuing an academic career leading to a degree in zoology and a doctorate in animal behaviour, specifically, polecats.

Over sixteen years were spent lecturing in physiology at Hull University with research being conducted in the fields of chemical ecology and hedgehog conservation, welfare and ecology. This resulted in several publications most of which are available on her website: www.tonibunnell.co.uk

In 1990 Toni Bunnell began running, single-handedly, York Hedgehog Rescue Centre, taking in dozens of hedgehogs each year that were in need of care and/or treatment. These included sick or injured animals and orphaned babies. The knowledge gained during the past 26 years of working on a daily basis with hedgehogs has helped greatly with the writing of this book. Toni continues to conduct research into various aspects of hedgehog ecology, including long-term monitoring of hedgehog populations in York using remote cameras.